Litterbugs come in every size

by
Norah Smaridge

illustrated by Charles Bracke

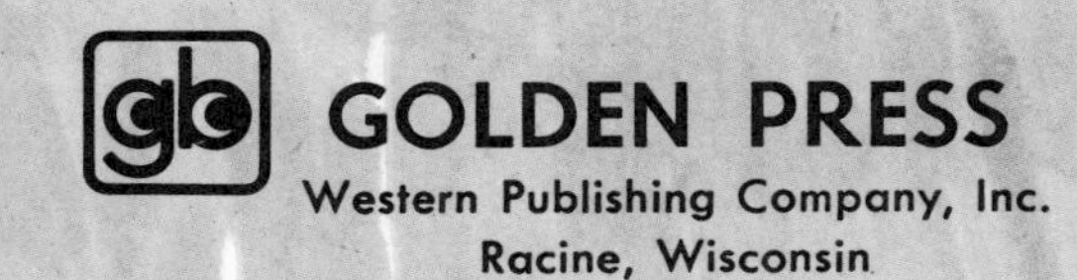

GOLDEN PRESS
Western Publishing Company, Inc.
Racine, Wisconsin

Fourth Printing, 1973

Litterbugs come
in every size.
(They're really monsters
in disguise!)

Sometimes they're tall,
sometimes they're teeny;

They're fat, or thin,
or in-betweeny.

Neatos, too, are every size,
But *they* are *angels* in disguise!
Everyone loves this helpful clan
(Especially the garbage man).

Whether they're yellow, white,
or black,
Litterbugs leave a messy track
Of empty cans and paper scraps,
Popsicle sticks and bottle caps.

The Neatos gather all this junk
And throw it in the trash can — PLUNK!

(Of course, it takes the snail forever,
But he's much better late than never!)

When Litterbugs
ride in a car,
They love to munch
a candy bar,
Then throw the wrapper
in the street,
Where it will stick
to someone's feet.

The Neatos leave things spick-and-span
(They use the nearest litter can).
They never drop a crumb, we've heard,
Except when asked to by a bird.

To Litterbugs,
our lakes and streams
Are only garbage dumps,
it seems,
For cans and paper
and — alas! —
Rusty knives
and broken glass.

Neatos keep water free from trash
So fish can swim and ducklings splash

And poodles paint and donkeys dream
Beside each lovely lake and stream.

Litterbugs make
lots of muss
On seats in parks
and on the bus.
They chew their gum
and leave a bit
Where some tired tiger
wants to sit.

The Neatos keep seats clean and neat;
They're careful where they put their feet.
Such thoughtfulness is kind indeed

(Especially from a centipede!).

When Litterbugs read books, we find,
They very often leave behind
A spot of milk, potato chip,
A bit of cake, or onion dip.

Neatos handle books with pride.
They never leave a crumb inside,
For books, they know, love you to read them
But much prefer that you *not* feed them.

When a Litterbug
has coughs and wheezes
And uses tissue
for his sneezes,
He crumples it
into a ball
And drops it —
anywhere at all!

The Neatos have a nicer plan:
Drop germy tissues in a can
And close the lid — for there's no doubt
This keeps those germs from popping out!

Neatos should hold a meeting soon
(Why not this very afternoon?)
And beg the Litterbugs to call
And then politely ask them all
If they are ready to repent
And clean up their EN-VI-RON-MENT.

And if the Litterbugs say no
(It isn't very likely, though),
All the Neatos then can do
Is throw *them* in the trash can, too!

But if the Litterbugs say yes,
The Neatos, full of happiness,
Can end the meeting with a bang
By asking them to join their gang!